Space Voyager
Uranus

by Vanessa Black

Bullfrog Books

Ideas for Parents and Teachers

Bullfrog Books let children practice reading informational text at the earliest reading levels. Repetition, familiar words, and photo labels support early readers.

Before Reading

- Discuss the cover photo. What does it tell them?

- Look at the picture glossary together. Read and discuss the words.

Read the Book

- "Walk" through the book and look at the photos. Let the child ask questions. Point out the photo labels.

- Read the book to the child, or have him or her read independently.

After Reading

- Prompt the child to think more. Ask: What are your favorite facts about Uranus?

Bullfrog Books are published by Jump!
5357 Penn Avenue South
Minneapolis, MN 55419
www.jumplibrary.com

Library of Congress Cataloging-in-Publication Data

Names: Black, Vanessa, 1973– author.
Title: Uranus / by Vanessa Black.
Description: Minneapolis, MN : Jump!, Inc., [2018]
Series: Space voyager
"Bullfrog Books are published by Jump!."
Audience: Ages 5–8. | Audience: K to grade 3.
Includes bibliographical references and index.
Identifiers: LCCN 2017028594 (print)
LCCN 2017033283 (ebook)
ISBN 9781624966927 (ebook)
ISBN 9781620318560 (hardcover : alk. paper)
ISBN 9781620318577 (pbk.)
Subjects: LCSH: Uranus (Planet)—Juvenile literature.
Classification: LCC QB681 (ebook)
LCC QB681 .B53 2017 (print) | DDC 523.47—dc23
LC record available at https://lccn.loc.gov/2017028594

Editor: Jenna Trnka
Book Designer: Molly Ballanger
Photo Researchers: Molly Ballanger & Jenna Trnka

Photo Credits: NASA images/Shutterstock, cover; Felix Mizioznikov/Shutterstock, 1 (child); Nostalgia for Infinity/Shutterstock, 1 (drawing); blackred/iStock, 3; dedek/Shutterstock, 4; JPL-Caltech/NASA, 5, 23bl; NASA, 6–7; Irina Dmitrienko/Alamy, 8–9; Ulga/Shutterstock, 10, 23tl; ImageDJ/Alamy, 11; Carlos Clarivan/Science Source, 12–13, 18–19, 23br; Mark Garlick/Science Photo Library/Getty, 14–15; Space Frontiers/Getty, 16; QAI Publishing/Getty, 17; Blend Images/SuperStock, 20–21; adventtr/iStock, 23tr; Rawpixel.com/Shutterstock, 24.

Printed in the United States of America at Corporate Graphics in North Mankato, Minnesota.

Table of Contents

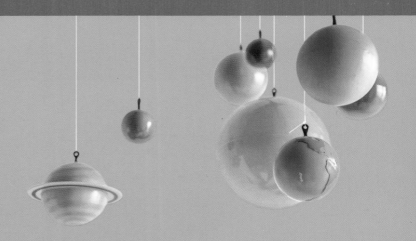

Big and Blue

What is the seventh planet from the sun?

It is big.

It is blue.

It is Uranus!

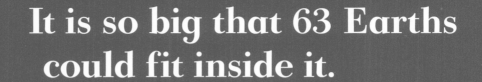

It is so big that 63 Earths could fit inside it.

It is an ice giant.

It is mostly liquids.

They are icy.

There are gases.
One is methane.

methane
gas

10

It gives Uranus
its blue color.

All planets spin.

But Uranus is different.

Why? It spins on its side.

Look at our solar system.

Uranus is far
from the sun.

It is very cold.

Uranus has 13 rings.
They are thin.

ring

moon

It has 27 moons.

How many spacecraft
have gone by it?

Only one.

There is more to learn.

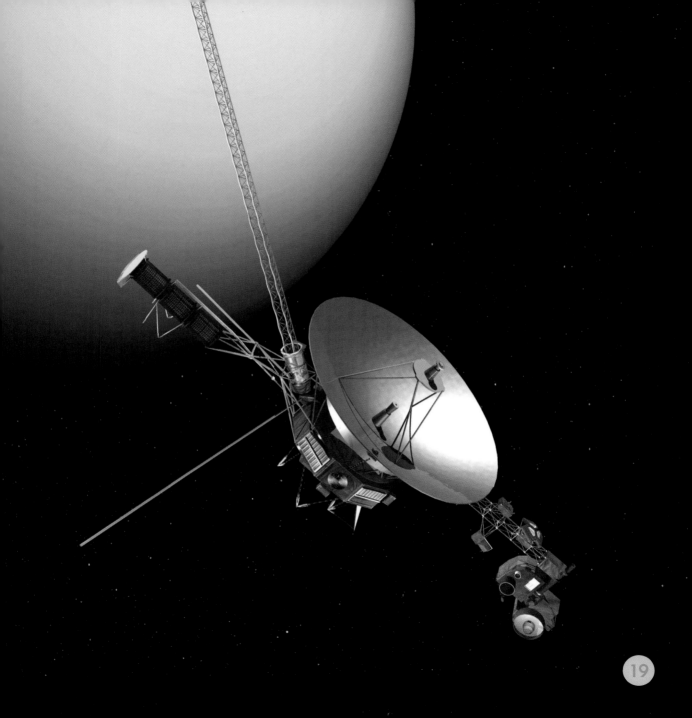

Wow!

What do you like about this planet?

21

A Look at Uranus

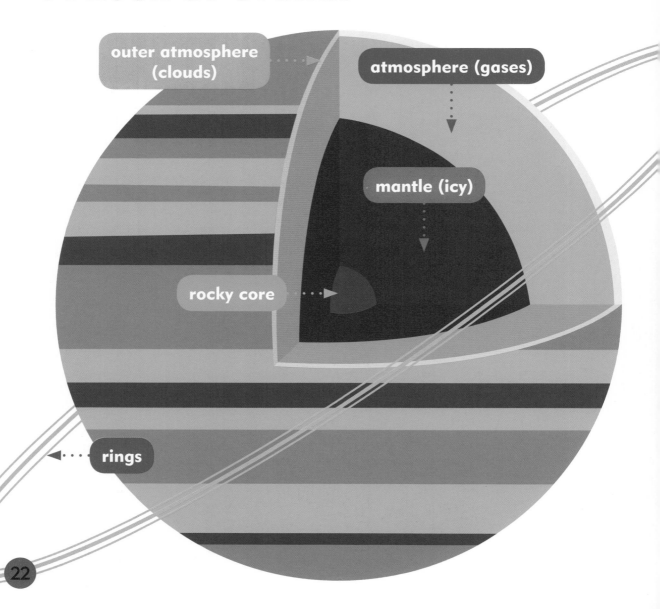

outer atmosphere (clouds)

atmosphere (gases)

mantle (icy)

rocky core

rings

Picture Glossary

methane
A kind of gas that gives Uranus its blue color.

solar system
The sun and other planets that revolve around it.

planet
A large body that orbits the sun.

spacecraft
Vehicles that travel in space.

Index

To Learn More

Learning more is as easy as 1, 2, 3.

1) Go to www.factsurfer.com

2) Enter "Uranus" into the search box.

3) Click the "Surf" button to see a list of websites.

With factsurfer.com, finding more information is just a click away.